BERRIES

Illustrations by
Hedvig Wright Ostern

BLANDFORD PRESS
Poole   Dorset

World Copyright © H. Aschehoug &
Co. (W. Nygaard), Oslo, Norway
First published in Norwegian in
1981 as Den Lille Baerboken by Eva
Maehre Lauritzen
English Text Copyright © 1981
Blandford Books Ltd, Link House,
West Street, Poole, Dorset BH15
1LL

ISBN 0 7137 1209 0

Typeset by CGM Graphics Ltd,
5b New Orchard, Poole, Dorset
BH15 1LY.

Printed by Mateu Cromo, Madrid

Introduction

Many berries are poisonous. These are marked with a 'P' in the text. Be particularly careful when dealing with poisonous berries found both in the wild and in gardens. Berries contain sugars, fruit acids, aromatic substances, vitamins and minerals. What are called berries in this book are really many different types of fruit: true berries, stone-fruits, false fruits, capsules and berry-like cones. Where the berry is botanically something other than a true berry this is indicated in the text.

The Blandford Mini-Guide to Berries will help to identify all the wild species found in the woods and fields.

Yew

Taxus baccata (P) Yew Family

The Yew is an evergreen like the spruce, Fir and Juniper. It grows on chalky soils, often in churchyards. The Yew's berry is a seed that develops with a red seed-coat (called an aril) around it. It looks very beautiful and tempting, but looks can deceive. Almost all of the Yew tree is poisonous. The only part that is not poisonous is the red seed-coat, but the seed inside is extremely poisonous. Nonetheless the Yew has been put to frequent use in folk medicine — often with serious consequences to the patient.

Even though the Yew is poisonous for people and livestock, birds are able to eat its berries without coming to any harm. The seeds pass through their digestive systems untouched and it is by this method of distribution that the species spreads itself. Yew trees can live to be extremely old.

Juniper

Juniperus communis Cypress Family

The Juniper originated in Ireland and is a native of the chalk downlands of Southern England, as well as being widely distributed throughout Europe.

Juniper berries are in fact small cones. The Juniper is usually 'dioecious' (having male and female flowers on separate plants). Berries are not, therefore, found on every bush. It will be noticed that both green and blue berries can be found on the same bush. This is due to the lengthy ripening period of the berries. One year after flowering the female bush produces berries, which take a further year to ripen and become blue. Meanwhile the bush has flowered yet again and produced additional green berries.

The Juniper is the oldest medicinal plant, and is also eaten with game and used in drinks, such as gin.

Bog Arum

Calla palustris (P) Arum Family

The Bog Arum grows in shallow
water, swamps and the margins of
bogs, and it flowers in early summer.
The small flowers are found grouped
together in corncob-shaped
inflorescences. The inflorescence is
surrounded by a shining white leaf
with the function of attracting insects
to the flowers so that they pollinate
them. Towards autumn, the Bog Arum
produces red berries that are tightly
packed together to form a cone-like
array. Now that the white leaf has
fulfilled its purpose it withers away.
The berries are filled with slime and
are poisonous, like the rest of the
plant.

Despite its poisonous nature, the
Bog Arum has been an important
fodder plant in certain areas. The
roots were collected and fed to pigs. In
time of great need the roots were also
used as human food.

Herb Paris

Paris quadrifolia (P) Lily Family

Herb Paris grows on good woodland leaf-mould. It is an easy plant to recognise with its characteristic whorls of four leaves.

The berries, which are formed at the tops of the plants, are bluish-black and are, to start with, surrounded by the withering flower. They taste unpleasant and are poisonous. They can easily be mistaken for bilberries, so great care must be taken when picking the latter.

In earlier times Herb Paris was known as the 'sore throat berry' in many country districts, because of its use as a cure for sore throats. The berries were also used in the dyeing of yarn. They gave a reddish colour which, unfortunately, was not colourfast.

May Lily

Maianthemum bifolium (P)
Lily Family

The May Lily is found only in some parts of Yorkshire, Lincolnshire and Middlesex, in both deciduous and evergreen woodland. It is quite rare. The plant has been called the 'May flower', because it flowers in May.

The small red berries form in a cluster in the autumn. The berries taste sweet but are poisonous and must not be eaten. (Unfortunately not everything in nature that is poisonous tastes unpleasant.) Some consider the berries of the May Lily to be so poisonous that there is some danger of very serious poisoning. Until the chemical contents of the berries have been further investigated, it is best to be careful and to take no chances. In any case the May Lily belongs to a group of plants, the convallarias, that contain many dangerous poisons.

Scented Solomon's Seal

Polygonatum odoratum (P)
Lily Family

Scented Solomon's Seal is the commonest of the three large wild convallarias and is also possibly the most poisonous. It has bluish-black berries and grows on high ground, stony ridges and in small woods.

The common Solomon's Seal *(Polygonatum multiflorum)* looks like a large scented Solomon's Seal, but in contrast to the usual arrangement for convallarias it has several flowers and berries associated with each leaf. Whorled Solomon's Seal *(Polygonatum verticillatum)* carries its leaves in whorls all the way up its stems. It grows in dense woodland. Unlike the two previously mentioned species it has red berries.

All of these three convallarias have poisonous berries.

15

Lily of the Valley

Convallaria majalis (P) Lily Family

Most people are familiar with the Lily of the Valley when it is in flower. Fewer people, however, have seen the Lily of the Valley in the autumn with its reddish-yellow berries. There are many reasons for this. In densely populated areas practically every single flowering plant is picked and they therefore never get the chance to form berries. Moreover the Lily of the Valley does not always form berries, even when it is left alone. This may be because it spreads itself effectively by means of subterranean runners.

The Lily of the Valley is an extremely poisonous plant. Like the previously mentioned convallarias, it contains a poison that affects heart function. It has a similar effect to *Digitalis*. The poison is found in the whole of the plant as well as in the berries.

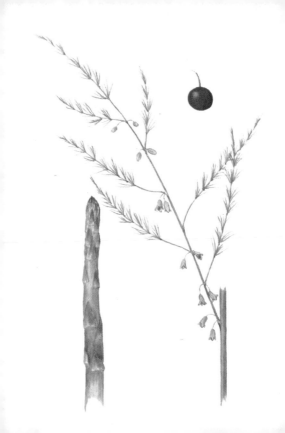

Asparagus

Asparagus officinalis (P) Lily Family

Asparagus grows in sand and on stony ridges by the sea. The part we eat is the white or pale green shoot which has had soil heaped up around it in order to keep it tender and well-flavoured. A long, branching, almost needle-shaped part of the plant comes up out of the earth. The Asparagus leaves, like small scales, are attached to this part of the plant. The reddish berries are treated as poisonous. Three black seeds are found in each berry. Birds eagerly eat the berries, thereby helping to spread the plant.

The Latin name of this plant, *Asparagus officinalis*, shows that it was once considered officinal, ie a plant used in medicine. It was used as a blood cleanser and diuretic. However, today it is used foremost as a vegetable.

Mistletoe

Viscum album (P) Mistletoe Family

Mistletoe is a small, parasitic bush that grows in the branches of certain species of tree, especially the lime, maple, apple and poplar. Mistletoe is green all year round, including the period after its host tree has dropped its leaves. The berries and the rest of the plant are poisonous and Mistletoe remains to this day a medicinal plant.

The berries of Mistletoe ripen in late autumn. They contain slime and a single seed. Birds eat the berries and spread the plant. In the past a sort of glue was made from Mistletoe berries in southern Europe. It was spread on the roosts of small birds, and the birds became immobilised in the thick glue.

It is an old English tradition to hang up Mistletoe at Christmas. Everyone who stands under the Mistletoe is entitled to a free kiss!

Baneberry

Actaea spicata (P) Buttercup Family

The Baneberry plant grows on good,
preferably slightly damp, soil rich in
leaf mould in woods and scrubland. Its
beautiful, black, shining berries look
tempting but have an unpleasant
smell and are poisonous. Fatal
poisonings have been recorded. Care
must be taken when handling the
plant, as it causes blistering on
contact with the skin. The plant was
used in the past as a medicine for
certain disorders. Its foremost use was
as a painkiller for arthritis and
toothache, but it was also used for
tonsillitis and sore throats. The Bane
berries were dried and ground and
spirits were poured over them.

Barberry

Berberis vulgaris Barberry Family

The Barberry has been introduced to
Great Britain and is now fairly
widespread, though never numerous.
These plants are relics of an ancient
medicinal plant that the Arabs
brought with them to Europe. Since
that time country after country has
declared war on this species of
Barberry, because it plays host to a
rust fungus, the black rust, that
attacks cereals and other grasses,
reducing harvests. Wild Barberry has,
as a consequence, been removed from
all areas containing agricultural land.
Several species of *Berberis,* however,
are popular garden plants.

The berries of the wild Barberry
taste sour but they are rich in vitamin
C. They have been used in some
countries to make juice and jelly.

Gooseberry

Ribes uva-crispa Currant Family

The Gooseberry grows wild in sunny spots. The berries on wild bushes are usually small and their colour can vary, being greenish, yellow or reddish. Some have smooth-skinned berries; others have hairy ones.

The Gooseberry is probably the oldest currant in cultivation, but it has not been treated as a cultivated plant for very long. Gooseberry-growing was a passion in many places, particularly in England, in the nineteenth century. A whole host of different varieties of Gooseberry were to be found in cultivation and some of these can still be found in gardens today.

At the beginning of this century, the Gooseberry was attacked by an American species of fungus. Today there are sprays that are effective against it, and there are also resistant varieties.

Blackcurrant

Ribes nigrum Currant Family

The Blackcurrant is one of the most popular garden berries. It is not at all easy to determine whether the apparently wild Blackcurrant bushes, which are found here and there, really are wild or are simply escapees from gardens.

The Blackcurrant truly a modern medicinal plant. Who has not tried Blackcurrant juice or Blackcurrant toddy when a cold or sore throat has threatened to take its toll?

Blackcurrants are rich in vitamin C and are, therefore, particularly valuable berries to eat fresh, in jams or as juice. Blackcurrants and Redcurrants have also been used for wine-making and are sometimes known as 'wineberries'. Blackcurrant leaves are also rich in vitamin C.

Redcurrant

Ribes rubrum Currant Family

The Redcurrant is found wild or
escaped from cultivation all over the
country. Redcurrant bushes can be
found in screes and on steep slopes
almost everywhere. In the past the
Redcurrant was an extremely popular
garden berry. The berries were eaten
fresh, or used for jam-making, for
juicing and in wine-making. The
berries are often very sour.

In many places it was common
practice to plant wild Redcurrant
bushes in the garden, as well as to
harvest wild Redcurrants. There
appear to be several closely related
varieties of wild Redcurrant, perhaps
three in this country, each with
different flowers and leaves.

Alpine Currant

Ribes alpinum Currant Family

The Gooseberry, Blackcurrant, Redcurrant and Alpine Currant all belong to the same genus. *Ribes.* Many decorative cultivated bushes such as the Flowering Currant and the Golden-leaved Currant also belong to this genus. They have in common palmately-veined leaves, and flowers and berries that are formed in clusters.

The Alpine Currant can be grown as a hedge plant, and it also grows wild. Its berries are so similar to those of the Redcurrant that they can almost be confused, but there are usually only 2–3 berries in a cluster. The taste, however, is very different. The Alpine Currant has a sweet, cloying taste and is best suited to decoration. The berries are not at all dangerous.

Bird Cherry

Prunus padus Rose Family

The Bird Cherry grows throughout Europe. It is light-loving and prefers to grow in open places, and is a very decorative tree. Despite the 30–40 flowers in each cluster, there are rarely many berries — often only 1–3 in each cluster. Its fruit sets very poorly.

The Bird Cherry, Wild Cherry, Dwarf Cherry and Blackthorn all belong to the same genus, *Prunus*. The Plum, Peach, Almond and Apricot also belong to this genus. They all have stone-fruits. Their stones are poisonous because they contain a bitter substance that can be split into prussic acid and grape sugar.

The shiny bluish-black Bird Cherries have a sweet, astringent taste that gives a strange, dry sensation in the mouth. They are usually eaten by children.

Wild Cherry

Prunus avium Rose Family

Botanists are uncertain whether Wild Cherries are truly wild or have simply spread from gardens. Wild Cherries are significantly smaller than cultivated cherries and are dark with a sweet, but slightly bitter, taste. The Wild Cherry has a long tradition as a cultivated tree. The oldest cultivated varieties have dark berries but many of the varieties now cultivated have lightly coloured flesh on their fruits.

Dwarf Cherry trees are usually smaller than Wild Cherries, and are often only bushes. Unlike the Wild Cherry, the Dwarf Cherry has leaves on its flower-stalks and its fruit-stalks. It also has smaller leaves. The Wild Cherry is best suited to being eaten fresh, and it is also used in the making of jams, juice, wines and liqueurs.

Blackthorn

Prunus spinosa Rose Family

The Blackthorn or Sloe is found throughout Europe. It is found on dry hills and at the edges of woods, often forming almost impenetrable thicket.

In the autumn the Blackthorn has extremely sour, bluish-black stone-fruits (often called sloes) with an astringent taste. They are like very small plums, and it is possible that the Blackthorn is one of the ancestors of today's cultivated plum.

There is considerable evidence that people were eating sloes as far back as the Stone Age. Sloe stones have been found at Stone Age dwelling sites in many areas. In some areas sloes are still used in the making of marmalade, wine and liqueurs. They should be exposed to a few frosty nights before being picked, as this makes them sweeter and gives them an improved flavour.

Cotoneaster

Cotoneaster integerrimus (P)
Rose Family

Cotoneaster bushes are small and
grow on sunny slopes, particularly
those with chalk as their bedrock. Its
flowers are slightly pink and rich in
nectar. They are arranged in groups of
a few flowers positioned where the
leaf-stalks meet the stems. The
alternately arranged leaves are
smooth on top but extremely hairy
underneath. In the autumn,
Cotoneaster has red berries which are
really small stone-fruits. They contain
a compound of prussic acid and are
poisonous.

Many other species of Cotoneaster
are cultivated in gardens, and are also
found growing in the wild, due to the
fact that their berries are widely
spread by birds.

Common Hawthorn

Crataegus monogyna Rose Family

The Hawthorn's berries are not, botanically speaking, true berries. The Hawthorn has a false fruit, a sort of miniature apple with a hard layer inside, known as a stone-apple. Stone-apples are not especially appetising. The flesh of the fruit is mealy, without a strong flavour.

The different species of Hawthorn are not easy to distinguish from each other. Some species grow wild or have become wild and many species are grown in gardens. They have white flowers, some with one style, others with two. The leaves have slightly different shapes. As the species can be hybridised, one can encounter intermediate forms that have arisen as a result of crosses between several species.

Crab-apple

Malus sylvestris Rose Family

The Crab-apple is found on mountainsides and at the edges of woods. Its apples are small and sour compared to garden apples. The wild Crab-apple has probably, over time, interbred to a certain extent with cultivated apples and there are undoubtedly one or two wild apple trees around that originated in gardens. Our garden apple and pear trees are grafted, whereas wild trees grow directly from apple and pear seeds.

The small sour Crab-apples have, since early times been picked and eaten. A little frost makes them sweeter. The sour apples give an excellent cider (sour garden apples can be used for this too). They were very popular in medieval times, and are still used to make jelly and wine.

Since ancient times the apple has been a symbol of love and power.

Wild Pear

Pyrus communis Rose Family

The commonest cultivated varieties of Pear trees are descended from Wild Pears, and in middle and eastern Europe Pear trees commonly grow wild.

The Wild Pear is a thorny tree or bush with small, hard pears at its top. The pears can be very hard and sour, but when allowed to ripen fully the taste improves. In countries where the Wild Pear is common, spirits and cider are made from them. A certain amount is also used as animal fodder.

Rowan

Sorbus aucuparia Rose Family

Rowan berries have been harvested
since ancient times. In bad times they
were used as human food but
ordinarily they were given to chickens
and to pigs and cows. The Rowan
berry is a false fruit with a structure
similar to that of the apple. Rowan
berries are not used very much today.
They hang and give pleasure to the
eyes as well as winter food to the birds.
Gourmets favour rowan berry jelly as
an accompaniment for game.

The Rowan berry contains a bitter
substance which can be rendered
milder if the berries are boiled and the
cooking water discarded. The bitter
substances are further camouflaged if
a little apple is mixed in. The Rowan
berry contains as much vitamin C as
the orange.

Whitebeams

Sorbus (several species) Rose Family

The Whitebeams belong to the same genus as the Rowan and share many features with it. The Whitebeams include a range of transitional forms from Rowan-like, almost pinnately-leaved forms to ones with complete, undivided leaves. The fruit of Whitebeams is similar in appearance to that of the Rowan and, again like the Rowan berry, is a stone-fruit. Most Whitebeams have mealy, inedible berries.

The Whitebeam shown in the picture is a Swedish Whitebeam *(Sorbus intermedia)* which is the most common of the species planted in gardens. To its side are shown leaves of the Common Whitebeam *(Sorbus aria)* which is one of the most common wild species.

Cloudberry

Rubus chamaemorus Rose Family

The Cloudberry is found especially in mountainous areas. The plants are particularly vulnerable during their flowering period. A promising Cloudberry harvest can easily be ruined by frost or a hailstorm during the critical period.

The Cloudberry plant is dioecious. The majority of the flowers yield berries. Cloudberries used to be important both in the diet and as a source of cash. Cloudberries are rich in vitamin C and preserve themselves well as they contain benzoic acid. They were used as a remedy for scurvy long before people knew what vitamins were.

Arctic Bramble

Rubus arcticus Rose Family

The Arctic Bramble is rare in Great Britain. It most closely resembles its relative the Rock Bramble, but its flowers are pink and its berries have a wonderful greenish-reddybrown colour. The Rock Bramble is occasionally used on the rock garden, because of its beautiful flowers.

The Arctic Bramble is considered to be the aristocrat of the berries. Its berries have a smell and flavour that make them unforgettable. Even where the Arctic Bramble is common, there can be some distance between berries. It is even more sensitive than the Cloudberry during its flowering period. Many years can pass between each bumper year.

Rock Bramble

Rubus saxatilis Rose Family

The Rock Bramble is common in stony
places, in woods and on moors. Like
the Cloudberry and Arctic Bramble,
the Rock Bramble, Blackberry,
Raspberry and Dewberry all belong to
the genus *Rubus*. The berries of these
plants have a characteristic structure.
They look like several small balls
assembled together. Each little ball is,
botanically speaking, a stone-fruit.
The berry is in effect an assembly of
stone-fruits that have developed
together.

It is not difficult to identify the
stones of the stone-fruits of the Rock
Bramble. They are quite large and are
only covered by a thin layer of
fruit-flesh. Usually only a few
stone-fruits go to form each Bramble
berry.

Raspberry

Rubus idaeus Rose Family

The Wild Raspberry is found in most of Great Britain. It tastes better than most cultivated Raspberries. Raspberries are delightful eaten fresh but they are also excellent for jam and for juice-making. The Raspberry is also better suited to freezing than most other types of berry.

One can often see luxuriant clumps of Raspberry bushes without a single berry. This results from the fact that the bush usually only forms berries on the previous year's shoots. Long branches rise up out of the earth from subterranean shoots. In the first year they are soft and herbaceous, without flowers; next year they form flowers and berries. Then they die and new shoots take over. The grub one often finds in the berries is the larva of the raspberry beetle. It is not dangerous to humans but can make Raspberries somewhat unappetising.

Dewberry

Rubus caesius Rose Family

The Dewberry is a small creeping bush that can easily be confused with the Blackberry. Its flowers are white and open, like those of a Blackberry, but the plant has many small thorns and is much smaller in size. The berries appear to be matt blue because of a light outer covering of wax.

The fruit of the Dewberry, like that of the Rock Bramble, most often consists of just a few little balls, each of which is a stone-fruit. Crosses (hybrids) between the Dewberry and the Rock Bramble, Raspberry or Blackberry occasionally arise in nature. These plants have characteristics of each of their parental stocks and make it even more difficult to distinguish the species from each other. Dewberries are safe to eat.

Blackberry

Rubus fruticosus Rose Family

The Blackberry is really the collective
name for a mass of different, closely
related species which even the
botanists find it difficult to
distinguish between. Most of them
have thorny shoots, big white or pink
flowers and almost black berries.

The berries have a fine, slightly sour
taste and are considered by many to be
a delicacy of the same order as
Cloudberries. The berries ripen first
late in the autumn. If they are to be
eaten fresh it is better to pick them
when they are slightly over-ripe.
Berries for jam-making can be
harvested a little earlier. When
picking Blackberries the flower
receptacles to which the fruit are
attached are often picked as well. The
berries have, therefore, to be cleaned
before use.

Wild Strawberry

Fragaria vesca Rose Family

The Wild Strawberry is a plant that most people know and that brings back happy summer memories for most of us. The Wild Strawberry, like the Wild Raspberry, has a flavour and a fragrance lacking in cultivated varieties.

It is best to pick the Wild Strawberry to eat fresh. According to an old saying, the first berries of the year should be trampled on straw. The Wild Strawberry was of greater importance to people in earlier days than it is today. There may well have been more of them in those days. Today many good Strawberry spots have been dug up or built on.

The Wild Strawberry is traditionally held to be health-giving and is particularly recommended for those who are anaemic.

Strawberry species

Fragaria species Rose Family

Fragaria viridis has berries that are
surrounded by long sepals that curve
over them. Its berries are, as a
consequence, more of a chore to pick.
They are more lightly coloured than
the Wild Strawberry and not as finely
flavoured. The first berries do not
usually ripen as early as the Wild
Strawberry.

The Hautbois Strawberry *(Fragaria
moschata)* is a species of strawberry
that has spread from gardens. Today
it is being increasingly displaced in
gardens by improved varieties of
cultivated strawberry. The Hautbois
Strawberry is stronger than the two
previously mentioned species. It has
raised flower-stalks with hairs that
stick out. It is dioecious, unlike the
other two species.

Wild Roses

Rosa (several species) Rose Family

All Roses have the same type of fruit
— hips. These are false fruits formed
when the flower base grows to form a
jug-like container in which are to be
found the true fruits, which are stones
or hairy seeds.

Hips can vary considerably in
appearance, be small or large,
spherical or elongated, smooth or
hairy. The colour can vary from
orange to deep red or almost black.

Hips are the fruit most rich in
vitamin C, having a higher content
than oranges. Hips are excellent for,
among other things, making jam, tea
and wine. All sorts of rosehips can be
used, whether wild or cultivated, but
their taste and quality will vary.

Holly

Ilex aquifolium (P) Holly Family

The Holly, with its dark green, prickly leaves, is quite unlike any other wild plant. In contrast to nearly all other deciduous trees it does not drop its leaves when winter comes.

The Holly is widespread in Great Britain as it thrives on mild winters and relatively warm summers. The Holly has small, white flowers formed in the crooks where leaf-stems meet the branches. The female bushes form decorative but poisonous red berries (that are really stone-fruits) in the autumn. Large quantities of Holly wreaths are sold at Christmas time. The tradition of bringing in Holly wreaths in the winter is far older then Christianity. It is a relic of ancient pagan fertility rite.

Spindle Tree

Euonymus europaeus (P)

The Spindle Tree looks rather modest when in blossom, with its long, narrow leaves and its inconspicuous yellowish-white flowers. It is only in the autumn when its seed capsules become red that this bush becomes spectacular. When the capsules break open, up to four orange seeds hanging on individual strings drop out. The orange seeds, which are similar to berries, have an orange outer layer called the seed coat.

The Spindle Tree is really a garden plant and several closely related species are cultivated. The berries and wood contain a particularly large quantity of poison. The name Spindle Tree comes from the past use of the wood in yarn spindles.

Alder Buckthorn

Rhamnus frangula (P)
Buckthorn Family

The bark of the Alder Buckthorn is an important drug as it contains strongly cathartic substances. It also contains poisons and is therefore either preserved for a long time or heated up before being used as a medicine. The poison is also found in the rest of the Alder Buckthorn.

The Alder Buckthorn is a bush or small tree that grows in thickets and deciduous woodland. It has bunches of small flowers where leaf-stems meet branches, is dioecious and has unserrated leaves. Its berries are initially green, later red and finally black. The berries do not all change colour simultaneously, so one can find branches with green, red and black berries at the same time. Alder Buckthorn berries are poisonous and are really stone-fruits with 2–3 stones.

Purging Buckthorn

Rhamnus catharticus (P)
Buckthorn Family

Purging Buckthorn is a bush or small
tree that grows on sunny slopes and in
small woods. The leaves, which come
in pairs, can, on fleeting inspection,
seem to be similar to those of the
apple. They are finely serrated along
their entire perimeter and their veins
form characteristic curves that run
parallel to the leaf-edge. The plant is
dioecious. It forms small bunches of
flowers at the junctions of the
leaf-stems and branches. The flowers
are small and greenish. The berries
are really stone-fruits. They are
poisonous and strongly cathartic.
They are green at first and black later.

Just like the *Berberis*, the Purging
Buckthorn acts as host plant for a
fungus that can attack cereals —
especially oats.

Daphne

Daphne mezereum (P) Daphne Family

Daphne is a very small bush which produces little red flowers on naked branches early in the spring. It grows on stony ground in woods, particularly where the bedrock contains calcium. It is very rare in Great Britain.

The entire bush is poisonous. If a twig is broken off and placed in the mouth, the mucous membrane will swell up and there will be a burning sensation in the mouth and throat. Daphne produces berries very early in the summer. These are red, tempting and very poisonous.

The bark and berries of Daphne are rich in traditional medicinal uses. Certain tracts of Daphne were famous for the external treatment of arthritis.

Sea Buckthorn

Hippophae rhamnoides
Elaeagnus Family

The Sea Buckthorn berry is a false
fruit formed by a calyx growing
around the true nut-fruit. The berries
are sour with a distinctive flavour,
which is not to everybody's liking.
They are difficult to gather because of
the bush's thorns and because it is not
easy to remove them from the
branches without damaging them. It
is very rich in vitamins. It is high in
vitamin A and vitamin C. The flavour
is best after a frosty night.

Sea Buckthorn is most common
around river mouths and beaches. The
bush sends out new shoots freely and
can form impenetrable thickets. The
wood was, in the past, used to make
rake teeth and the berries were used to
a certain extent. They can be used to
make juice, jelly and liqueurs, and can
be mixed with other fruit to make jam.

Dwarf Cornel

Cornus suecica Dogwood Family

The large, bright red berries of the
Dwarf Cornel are delicate and
tempting, but their sweet, distasteful
flavour makes them unsuitable for
eating. They have also earned the
completely unwarranted reputation of
being poisonous. The berries are
stone-fruits with two seeds.

Dwarf Cornel berries are
exceptional in having almost black
flowers — a rarity in the plant
kingdom. The flowerheads are
surrounded by four white leaves that
most people take to be petals. These
four leaves sometimes encircle as
many as twenty-five of the small
blackish flowers. The white leaves
give the illusion that the entire
flowerhead is just one flower.

The Dwarf Cornel is found mostly in
Scotland, but it is never common in
Great Britain.

Common Dogwood

Cornus sanguinea Dogwood Family

The Common Dogwod is also called 'hornwood', on account of its hard hornlike wood. This bush is up to 2 metres high and has intense withered bark and leaves arranged in pairs. In the spring it has dense umbellate clusters of white flowers and in the autumn it has bluish-black stone-fruits with an unpleasant bitter flavour.

The Common Dogwood forms quite large bushes in small woods. The smooth red bark means that it cannot be confused with any other wild bush. Unlike the Common Dogwood, the White Cornel *(Cornus alba)* has white or pale blue berries. Its leaves have several side veins and are longer and more pointed. Several other species of Dogwood are also cultivated in the garden.

Ivy

Hedera helix (P) Ivy Family

The Ivy is an evergreen climbing
shrub. It can form tight masses
amongst rocks and scree or wind its
way around trees. Ivy thrives on mild
winters and is common in Great
Britain.

The flowering period of Ivy is from
September to October and further into
autumn. Its modest yellow flowers are
grouped in umbels. Its fruits are
berries that ripen in the following year
in the spring. Both the berries and the
rest of the plant are poisonous. The
berries are spread by birds.

When the Ivy germinates to form a
new plant, its first leaves are palmate
like those of a maple. After a few years
it begins to put out shoots with entire
pinnately-veined leaves. It is only
these shoots that can flower and
produce berries.

Alpine Bearberry

Arctostaphylos alpina
Heather Family

Unlike its near relative the Bearberry, the Alpine Bearberry drops its leaves every autumn. Before doing so, the whole plant displays fine autumn colours. It is at this time of year that it is easiest to find the Alpine Bearberry.

The Alpine Bearberry is local to remote highlands in Great Britain. The plant thrives where it is well exposed to sunlight (ie where there are no trees or other objects that cast shadows).

The fruit of the Alpine Bearberry has the reputation of being poisonous. There is no evidence, however, that the berries do contain poisons. The green berries, which are really stone-fruits, become red, before they gradually turn to a shiny black. They are about the same size as a Blackcurrant and have a somewhat sweet, cloying flavour.

Bearberry

Arctostaphylos uva-ursi
Heather Family

The Bearberry plant can be confused
with the Cowberry plant, and its red
stone-fruits are, at first glance, very
like those of the Cowberry.

The Bearberry, like the Cowberry, is
an evergreen dwarf bush. The
Bearberry leaf has a flat edge,
whereas that of the Cowberry is bent
over. A delicate network of veins is
found on the underside of the
Bearberry leaf, whereas the Cowberry
has dark dots. The Bearberry is
completely useless as a domestic
berry. The plant's berries are dry and
floury, and not, as many believe,
poisonous.

The Bearberry has been, and to a
certain extent still is, an important
medicinal plant. It is used, in
particular, in the preparation of
medicaments for complaints of the
urinary tract.

Cowberry

Vaccinium vitis-idea Heather Family

The Cowberry grows in well lit places where heather grows. It is common throughout the country. Sometimes there are so many Cowberries in clearings and on heather moors that the landscape glows red in the autumn.

The Cowberry is not particularly rich in vitamins but it contains benzoic acid which makes it easy to preserve, even without sugar. This was especially important in the past, when sugar was either hard to get or an expensive luxury. Cowberries are used mainly to make a jam which, in some countries, is a regular accompaniment to nearly all forms of dinner food.

The Cowberry also has a past history as a medicinal plant. Tea made from Cowberry leaves has been used as a cure for colds.

Bog Whortleberry

Vaccinium uliginosum
Heather Family

The Bog Whortleberry is a close
relative of the Bilberry. It is local to
Scotland and North England, and
grows at the edge of bogs and in
humid woods.

The Bog Whortleberry plant is
usually taller and stronger than the
Bilberry plant. The leaves are leathery
and have a characteristic blue-green
colour on their undersides. Bog
Whortleberries are a little larger than
Bilberries and have a layer of wax on
the outside that makes them lighter in
colour. The berries are pale inside and
their juice is colourless. The Bog
Whortleberry plant has beautiful
autumn colours.

Bog Whortleberries lend themselves
best to jam-making, but they can also
be used to make juice and wine. The
berries contain at least three times as
much vitamin C as the Bilberry.

Bilberry

Vaccinium myrtillus Heather Family

The Bilberry is a common and popular plant in Great Britain which is sometimes eaten with pancakes. It can be eaten fresh as well as frozen or in the form of jam or juice. Bilberries can also be used to make wine, their colour making them particularly suitable for red wines.

The Bilberry is not especially rich in vitamins but contains plenty of sugar and aromatic substances which make it so well flavoured. The berries also contain tannins which is what probably makes them so effective against diarrhoea and upset stomachs. Dried Bilberries are particularly effective for these uses. The Bilberry is in fact an ancient medicinal plant and is still of some medical importance.

Cranberry

Vaccinium oxycoccos Heather Family

The Cranberry is a creeping dwarf which, at most, rises 5 cm above the ground. It often lies well hidden in the moss of peat bogs.

The berries ripen in the late autumn and are extremely sour. After a frosty night the berries become translucent and sweeter in taste. The berries are extremely rich in vitamin C. They can be used to make jam, jellies and juice. The Cranberry contains benzoic acid which means that the fruit preserves itself well. Cranberry jam can be used in much the same way as Cowberry jam.

In addition to the ordinary Cranberry, there is also the Small Cranberry.

Crowberry

Empetrum nigrum Crowberry Family

The Common Crowberry is dioecious, and fruiting can be uneven. The mountain Crowberry *(Empetrum hermaphroditum)* grows in mountainous regions. It is monoecious (male and female flowers on the same plant) and therefore has more reliable pollination and fruiting. Many consider that it has far better berries than the other species.

Crowberries are really small stone-fruits full of juice and containing many stones. They have been used to make juice, dyestuffs and ink. The Crowberry is used as a diuretic in folk medicine. Berries preserved in spirits are a traditional drink, with a fine red colour. Crowberry twigs were reputed to make particularly good brooms and brushes.

Privet

Ligustrum vulgare (P) Olive Family

The Privet is common in Great
Britain, and this and other species of
the genus *Ligustrum* are planted in
gardens. They are often used as
hedges, as they tolerate cutting well.

Privet has shiny, green leaves, most
of which stay on the bush throughout
the winter. The pale yellow flowers
form clusters. They have a slightly
unpleasant smell. The bushes can be
up to 2 m high. From time to time they
put out strong new shoots which
means that they form tightly packed
bushes.

The black berries ripen in
September–October. The berries are
very poisonous. The bark and leaves
contain, in addition to the poison, a
substance that induces eczema and
acts as a skin irritant.

Bitter Sweet

Solanum dulcamera (P)
Nightshade Family

Bitter Sweet belongs to the same genus as the potato — a genus containing many poisonous plants. The potato is no exception: the whole plant is poisonous except for the tubers, which are the potatoes we eat.

Bitter Sweet is a creeping shrub-like plant. It look almost herbaceous, but the lower part of its stem is woody. Its violet flowers are a reminder of its kinship with the potato. Bitter Sweet thrives best in damp places but is not uncommon in small woods and sides of roads.

Its berries are red and elongated. They ripen gradually, often while the plant is still flowering, and are poisonous, as is the rest of the plant. Bitter Sweet has been used by both folk and orthodox medicine in the treatment of both people and animals.

Black Nightshade

Solanum nigrum (P)
Nightshade Family

The Black Nightshade is common along the sides of roads and in gardens, but can also be encountered as a beach plant. This plant has spread from Europe to the rest of the world other than the polar regions. The Black Nightshade is a perennial in warm countries; in colder regions it is an annual.

The Black Nightshade has spherical, black berries. These, like every other part of the plant, are poisonous. Some observations suggest that when the berry is fully ripe it no longer contains poison, but an alternative explanation is that people vary in their tolerance of the poison. It is cultivated as a crop in tropical regions. It is used as a leaf vegetable in much the same way as spinach.

Box Thorn

Lycium barbarum (P)
Nightshade Family

The Box Thorn has thin, thorny branches which bend downwards. The bush is approximately as high as a man. It can put out subterranean runners which can lead to a single bush turning into a thicket in the space of a few years. Its flowers come in groups and are positioned where the leaf stems meet branches. They are reminiscent of potato flowers.

Box Thorn berries are golden red and elongated. They are poisonous but probably not to such a degree as the previously mentioned species of the nightshade family.

Box Thorn can be grown near coasts, as it is not damaged by sea water.

Honeysuckle

Lonicera periclymenum (P)
Honeysuckle Family

The thin stems of the Honeysuckle wind their way around trees throughout Great Britain. Any tree that supports the Honeysuckle may find after a few years that it is being destroyed by it. As the tree grows in diameter, the grip of the Honeysuckle becomes tighter and tighter and the tree may be strangled.

The yellow flowers of the Honeysuckle proclaim its kinship with the cultivated Honeysuckle and other garden shrubs. Its flowers emit an intense odour, especially in the evening. They are rich in nectar and are frequently pollinated by night-flying moths. Both the smell and the pale flowers are specially designed to attract them. All of the Honeysuckle is poisonous — including the red berries.

Woody Honeysuckle

Lonicera xylosteum (P)
Honeysuckle Family

The Woody Honeysuckle has far less spectacular flowers than its relative the Honeysuckle. They come in pairs and are smaller and paler and completely odourless. It is fairly rare in Great Britain.

Its red berries, which ripen in August–September, are highly poisonous. Many of the Honeysuckles grown in gardens are also carriers of poisonous berries. Children are especially liable to be poisoned, as the berries are sweet. The branches of the Woody Honeysuckle have a loose, porous pith that is easily pressed out. Lengths of stem were, therefore, used to make pipe stems. In many places the Woody Honeysuckle was also used as a cure for arthritis.

Blue-berried Honeysuckle

Lonicera caerulea (P)
Honeysuckle Family

The berries of the Blue-berried Honeysuckle are black with a coating of bluish wax. They are poisonous to people but are eaten freely by birds. who thereby spread the bush. This plant is grown as a garden plant in Norway, as it is one of the hardiest and most pleasing plants to grow.

The leaves of the Blue-berried Honeysuckle lie in pairs and are smooth with blue-green undersides. The pale yellow flowers are formed in pairs, each pair forming one berry-fruit. It is possible that the Blue-berried Honeysuckle and one or two other related garden species contain, in addition to the usual Honeysuckle poison (xylostein), a further poison that has yet to be identified.

Red-berried Elder

Sambucus racemosa (P)
Honeysuckle Family

The Red-berried Elder was introduced
to Great Britain, and is now quite
common in Scotland. It has large,
pinnate leaves. Its yellowish-green
flowers hang in clusters, The coral red
berries are already ripe by July and it
is then that the bush is most readily
noticed. The red berry is really a
stone-fruit with three stones. The fruit
is poisonous before it is ripe. The
fruit's flesh contains no poisons after
ripening but the stones are still
poisonous. So too are its bark and
leaves. It is debatable whether the
fruit of the Red-berried Elder can be
eaten or used in any way in the
household. Some experts discourage
all use of the Red-berried Elder,
whereas others simply advise the
removal of the stones.

Common Elder

Sambucus nigra (P)
Honeysuckle Family

The Common Elder is common to
Great Britain and is prized as one of
the most delightful trees of our
countryside. It flowers around the
middle of the summer and later bends
under the weight of the hanging
clusters of Elderberries. The Elder
prefers deep fertile soil and a place in
the sunshine.

The white flowers of the Elder smell
strongly and hang in large, flat
umbels. Everywhere that the Common
Elder is found, it has been customary
to make tea with the flowers. This was
said to help with diseases of the
throat, colds and such like. The leaves
and unripe fruit are poisonous. When
the blue-black fruits are ripe they are
free of poison. The fruit is used to
make Elderberry juice and Elderberry
wine.

Guelder Rose

Viburnum opulus (P)
Honeysuckle Family

The Guelder Rose is most easily
recognised by its characteristic white
flowers, grouped in umbellate clusters.
The flowers around the outer edges of
the clusters are large and
eye-catching, whereas those towards
the centre are small and less striking
to the eye. The outer flowers are in fact
sterile and merely serve to entice
insects to the plant in order to
pollinate the more central flowers.
These smaller, more modest inner
flowers develop normally and form
fine, red berries that are really
stone-fruits.

The palmate, opposed leaves of the
Guelder Rose are somewhat similar to
the leaves of the maple. The red berries
ripen in September. The berries, bark
and leaves contain a poison.

Bryony

Bryonia alba (P) Gourd Family

Bryony is monoecious. It has black berries. The Red Bryony *(Bryonia dioica)* is dioecious and has red berries. Both species are old medicinal plants that contain poisons. The berries are poisonous. In the days when Bryony berries were used as medicine it was not unknown for patients to be poisoned as a result.

Once Bryony starts growing as a weed it is not readily removed. All the parts above the ground wither away in the autumn on the first frosty night but the roots survive and put out new shoots in the following spring. Despite the fact that the plants are poisonous, Bryonies have been held to be bringers of good luck.

Index of Common Names

Index of Scientific Names